Joseph-Eric Nnomenko'o

Extraterritorial Land Tenure and the Legal-Ethical Framework

Joseph-Eric Nnomenko'o

Extraterritorial Land Tenure and the Legal-Ethical Framework

China in Sub-Saharan Africa

ScienciaScripts

"We look at the African continent as if we were looking at the beds of sick people who are only talked to about their illness, as if they no longer existed as people: pathology takes precedence over all other personality or character traits. Just as the patient is reduced to the disease that plagues him, Africa is all too often reduced to its ailments. To make matters worse, in both cases, doctors and nurses often discuss diagnoses and remedies amongst themselves, without any concern for involving the patient. The simple question "Where does it hurt?" is barely asked by the curators, who already have their own ideas on the subject. This idea is all the more entrenched as they are not always strangers to the ailment and its evolution".

Anne-Cécile Robert

A
Alleluia Ministries International

Table of contents

General introduction

Adopting the logic of Karl Popper's critical rationalism[5] , this book follows in the footsteps of works dealing with the issue of offshore land concessions in sub-Saharan Africa. However, rather than carrying out a synthetic study based on the many apriori views of China-Africa cooperation in all its aspects, we have decided to focus on the Chinese presence in agricultural land, particularly in Cameroon, with a view to a detailed analysis of the accusations of land grabbing conveyed by certain international media and NGOs, which present a truncated or even false cartography of the Chinese presence in agricultural land in sub-Saharan Africa in general and in Cameroon in particular.

A number of researchers, whose work on China-Africa cooperation is both relevant and scientific, unfortunately continue to perpetuate, wittingly or unwittingly, the idea of illegal access to land by China in sub-Saharan Africa, using the expression "land grabbing". This expression is confusing for all concerned, and raises questions about China's methods, modes and even procedures for accessing land in sub-Saharan Africa.

Accusations of land grabbing by China in sub-Saharan Africa are problematic, especially as China's presence in African agricultural land remains minimal. Rather, such land acquisitions are a springboard to other sectors, such as the financing of activities in the urban infrastructure sector, particularly in the resource and raw materials exploitation sector (Gabas, 2013).

There is often confusion between the intention to acquire Chinese land, and the actual acquisition. One of the most striking examples concerns the famous 60,000 ha of land that China wanted to acquire in Senegal for sesame production, a project that remained a cognitive representation of reality. After realizing that the project could not operate on ten hectares, the Chinese investor finally withdrew. However, a number of media outlets and NGOs have incomprehensibly remained focused on a Chinese land grab of 60,000 ha

---

[5]Critical rationalism is an epistemological concept defended in particular by Karl Popper, according to which it is impossible to provide a certain foundation for scientific knowledge. Scientific knowledge is considered to be conjecture, the validity of which is constantly subject to criticism or refutation.

in Senegal, even though nothing has been .[6][7]

China's accusations of land grabbing raise the question of how large-scale assets are acquired, a matter of little public knowledge due to their confidential nature. This makes it difficult to distinguish between the announcement of a project and its actual implementation. While some operations involving hundreds of thousands of hectares remain little-known and receive little media coverage, others involving modest areas are surprisingly well publicized, such as those highlighted in the World Bank's September 2010 report on Nigeria and Ethiopia. This can certainly be explained by the fact that these projects generally involve local investors. However, governments sometimes call on foreign investors directly for large-scale operations. For example, in August 2008, Sudan published the following advertisement in *the Financial Times*: "State gives away 880,000 hectares of arable land "3.

To avoid the confusion that generally reigns between intentions to acquire land, fanciful figures and much-publicized accusations of Chinese land grabbing, it's important to be able to distinguish between the announcement of an acquisition or the actual acquisition (covering a much smaller area), and the start of operations (even smaller) on scales that are quite different from one another. Eleven different reference levels need to be taken into account in order to know what figure we're talking about:

- The global *investment area* offered by the host country to the investor,
- *Priority* investment areas,
- *Negotiated* areas,
- Options on future concessions *(Optional investment area)*,
- Leased *areas,*
- *Reserved land area,*
- Project areas,
- *Irrigated* or agro-industrial perimeters *(Irrigated area; Pivot area)*,
- Experimental areas,
- Industrial areas,

---

[6]*Ibid,* p. 17.
[7]Jean-Paul Charvet, *"Land grabbing"*, *Encyclopaedia universalis* [online], accessed February 25, 2023-URL : http//www.universalis-edu.com/encyclopédie/land land grabbing.

- *Resettlement* areas.[8]

If these different levels of reference were often taken into account in more sententious than objective studies of alleged Chinese land grabs in Africa, they would certainly help to avoid any accusations of Chinese land grabs on the macro scale of sub-Saharan Africa, and particularly on the micro scale of Cameroon, which is the focus of this study.

As in most sub-Saharan African countries, China is accused of land grabbing in Cameroon. In the November 14, 2010 issue of *Enquête Exclusive*, broadcast by the private French TV channel M6, journalist Bernard de la Villadière, not to be named, describes a Cameroon that is going through hell because the Chinese are stealing the work of Cameroonians. The journalist goes on to say that the Chinese have already monopolized thousands of hectares of land to produce rice, which is then shipped back to China. The documentary even has Cameroonians saying that the worst part of all this is that the best rice is sent to China, and only the broken rice is left in Cameroon.

In the same vein, in an article entitled *Quand le Cameroun nourrit la Chine (When Cameroon feeds China)* published on October 21, 2010, by the French weekly Politis.fr, journalist Simon Gouin reports that China is reportedly grabbing land in Cameroon to feed its huge population. According to the journalist, China is engaging in this practice because its agricultural model is based on imports.

In a documentary entitled *Razzia chinoise sur les terres camerounaises (Chinese Razzia on Cameroonian Land)* broadcast in 2009 by the French television channel ARTE, journalist Patrick Fandio asserts that the Chinese group IKO has set out to conquer Cameroonian farmland with the blessing and support of the highest levels of government in Beijing. Continuing his argument, the journalist asserts that in Cameroon, the Chinese could eventually control the country's entire cereal production chain if nothing is done. [*] Another article published by the NGO GRAIN entitled *Unmasking the land grab in Cameroon by a Chinese company*[9] denounces Chinese land grabs in Cameroon via its company Sino-Cam Iko Ltd. The NGO claims that China has already grabbed huge

---

[8] Gérard Chouquer, "Understanding massive land acquisitions in the world today, " www.fig.net/pub/fig2012/papers/its01h/TSO1H/ chouquer 5932.pdf. Accessed on 11/05/2024

[9] Cf.http://farmlandgrab.org/post/view/16667. Accessed on 16/05/2023.

hectares of land in the department of Haute-Sanaga.

As we can see, the above-mentioned media and NGOs support the idea of a Chinese land grab in Cameroon. Drawing on Émile Durkheim's socio-anthropological approach, the book suggests that we should be careful about the words we use to describe certain realities. For some trivially used words very often fail to convey the reality of the things to which they refer. Words, we believe, must correspond to what they are used for. From a strictly sociological point of view, ÉmileDurkheim advises us to be wary of vulgar notions and words that we use with confidence, as if they corresponded to things that are well known and defined, when in fact they only awaken in us confused notions, prejudices, passions and mechanistic interpretations(Durkheim, 2007).

The expression "land grab" fits into this category of vulgar notions. The Durkheimian approach is an invitation to always establish a relationship between the word and the reality alluded to, with a view to avoiding confusion between scientific investigation, which aims to establish **the veracity of facts, and inquisition, which very often tends to formulate fallacious and sententious judgments.**

In order to avoid the clichés and preconceived or even caricatured ideas that abound about China-Africa, this book adopts a circumspect and cautious approach, aimed at highlighting the gaps in certain information about Chinese agricultural activity in sub-Saharan Africa, and particularly in Cameroon. For it's always easy to take shelter behind the quotations of those who have gone before us, without seeking to verify the robustness of their writings. And yet, whether we like it or not, the failure to verify such information contributes to the propagation of rumours and the peddling of rumours which, by dint of being printed, become commonplaces and self-evident facts that are never questioned (Gabas and Chaponnière, 2012).

For example, we know that China has large chicken farms in Zambia, that the Chinese scare Africans[10] , that there were anti-Chinese riots in Yaoundé in 2008[11] ... that the Chinese abound in Africa.

---

[10] Jacques Barrat, "La CHINAFRIQUE : Un tigre de papier ", *Géostratégie,* n°25, 2009, p. 162.
[11] *Ibid,* p.162.

Chapter 1: China-Africa media

Introduction

In the space of just a few years, Africa, with its immense subsoil riches, has become the preferred location for numerous foreign investors, notably China and Western countries, who are waging an unprecedented economic and strategic war against each other. As in all wars, each player uses the weapons at his disposal to gain better control of the situation. In this confrontational competition between China and the major Western powers, the media are particularly hard at work. It's hardly surprising, then, that most information on China-Africa is false. This attitude is part and parcel of consciousness manipulation, disinformation, media hype and media spectacle.

1.1 Conceptual clarifications

Médiaspectacle

The media spectacle is the well-known method of media diversion associated with advertising hype and audiovisual non-culture. The main way to save time is to distract people, making them take sides over trivialities. This keeps the general public's attention away from the fundamental issues relating to the extreme seriousness of the current situation .[12]

Disinformation

Disinformation refers to a communication process in which the media are used to transmit partially erroneous information, with the aim of misleading or influencing public opinion to act in a certain direction. This is organized lying. Disinformation comes from the repeated bias in the selection of facts and the way they are angled (Le Gallou, 2013).

Disinformation is a conscious choice, a chosen, deliberate and planned approach to

---

[12]Sylvain Timsit, "stratégies de manipulation: les stratégies et les techniques des maitres du monde pour la manipulation de l'opinion publique et de la société, www.Syti.net/Manipulation.html, accessed 06/11//2020

manipulating public opinion and gaining the support of individuals and groups for an ideology and the actions that flow from it. The problem here is not so much information as information processing; it's a deliberate political choice to manipulate information in order to divert public opinion .[13]

1.2. Manipulation strategies

In his article entitled *stratégies de manipulation : les stratégies et les techniques des maîtres du monde pour la manipulation de l'opinion publique et de la société* , Sylvain Timsit (2002) lists ten strategies for manipulating public opinion and society. Here are just a few of them:

The diversion strategy

It consists in diverting the public's attention from the important issues and changes decided by the political and economic elites, through a continual deluge of distractions and insignificant information.

Addressing the public as young children

A communication strategy that uses discourse, arguments and a particularly infantilizing tone, often close to that of a debater, as if the viewer were a toddler or a mentally handicapped person.

Appeal to the emotional rather than the reflective

It's a classic technique for short-circuiting rational analysis, and therefore the critical faculties of individuals.

Keeping the public ignorant and stupid

The technique of making the public incapable of understanding the technologies and methods used to control and enslave them .[14]

---

[13]Boulan, "Chine-Afrique : La grande désinformation", www.mboaconnected.com/articles/.../ chinafrique-la- grande-désinformation. Accessed on 21/02/2020

[14]Sylvain Timsit, "stratégies de manipulation: les stratégies et les techniques des maitres du monde pour

It is these technical mechanisms for constructing a manipulative message that reveal a dual concern: to identify the resistance that might be put up against them, and to mask the very approach used by a certain business press to create confusion around the alleged Chinese land grabs in sub-Saharan Africa.

1.3. Some theses on China-Africa

Le Gallou's thesis

According to Le Gallou (2013), the information that is served up on China-Africa is prepared, oriented, planned so as to reach consciences and shape opinions. In absolute terms, there is no information. It's a fact that the media decide to bring to their audience's attention by presenting it from a certain angle. Concluding his analysis, Le Gallou argues that facts about China are presented with a deliberate bias, having been filtered and made to conform to editorial lines not always favorable to the interests of African populations [15].

Pougala's thesis

In his article *Les plus gros mensonges sur la coopération entre la Chine et l'Afrique (The Biggest Lies about Cooperation between China and Africa),* Jean-Paul Pougala refers to a curious anti-Chinese campaign organized in Britain's main cities, using posters and giant billboards in airports, Metro stations and crossroads. The campaign was paid for by one of Britain's leading news publications, The Economist. Headline: *Booming Chinese investments in Africa is bad for africans.* The poster explains why Chinese investments in Africa are terrifying for Africans for three reasons: a) the Chinese support dictatorial governments in Africa; b) Chinese garment factories in South Africa pay less than the minimum wage; c) elephants are a threat to Africa's livelihoods.

are disappearing in East Africa because of the Chinese. At the end of the article, the British public was asked which side they wished to take. Exasperated by this misleading publicity, the author ironically pointed out that the famous British magazine had forgotten

---

la manipulation de l'opinion publique et de la société", *Ibid.*

[15] Jacques-Yves Le Gallou, quoted by Boulan. In " Chine-Afrique : la grande désinformation ", *Op. cit.p.6*

to add that a) China is responsible for the torrential rains and mosquitoes in Africa; b) China is responsible for the lack of water in the Sahara and Kalahari deserts.

Certain international media have been feeding international public opinion with false information about Africa for years. Africans are often the first to be astonished by what these media say about Africa. The media's lack of knowledge about Africa is dramatic (Robert, 2004). This is what leads them to convey dangerous prejudices, such as the land grabs that are allegedly taking place there.

Africa's dramatic economic and social situation is distorting the way we see Africa, and the way it sees itself. The continent is increasingly identified with its own ills. It is no longer reduced to a single problem or series of problems: famine, war, AIDS and other pandemics, misery, ecological disasters, etc. This vision, which obviously corresponds to reality, tends to make Africa a vampire: it exists little or not at all as a place of life, of social and cultural expression. It is still not the subject of its own destiny, but a matter of concern. It does not exist as a free actor in the world, but as a performer in a sinister pantomime concocted by *others.*[16]

Conclusion

Africa offers enormous opportunities, with the varied riches of its subsoil exerting an undeniable attraction on the various economic players jostling for position. Under these conditions, no one wants to be left behind. From emerging to developed countries, everyone is strategically positioning themselves to avoid being left behind.

---

[16] Anne-Cécile Robert, *L'Afrique au Secours de l'Occident,* Paris, Edition de l'Atelier et Ouvrières, p.71

Chapter 2: Structuring land tenure as a legal-political and legal-commercial code in Cameroon

Introduction

Land tenure is the essential element in the organization of domanial, cadastral and land tenure systems. It is the relationship defined by law or custom with respect to land. It is a set of rules drawn up by a society to govern the behavior of its members with regard to the use of land resources. In other words, the distribution of property rights, the allocation of rights to use, control and transfer land, and the corresponding responsibilities and limitations. Land tenure is an important element of social, political and economic structures. Land tenure is multidimensional, involving social, technical, economic, institutional, legal and political factors.

In Cameroon, land management is organized through the colonial decree of July 21, 1932, which instituted the land registration system in Cameroon, and through ordinances no. 74/1 and 74/2 of July 06, 1974, which set out the land tenure system and the state tenure system. This organization involves the rational distribution of land, the resolution of land disputes and the repression of infringements of other people's land ownership.

2.1. Cameroon's land tenure system is organized around four areas

Private property

The State guarantees the right of all individuals and legal entities owning land to enjoy and dispose of it freely. As custodian of all land, the State may, in this capacity, intervene to ensure its rational use or to take account of the imperatives of defending the nation's economic opinions. The conditions for such intervention are set by decree. [ord. n 74-1 du 6 juillet 1974 fixant le régime foncier] The following lands are subject to private property rights: a) registered lands; b) freehold lands; c) lands acquired under the transcription system; d) definitive state concessions; e) lands consigned to the Grundbuch (Art 2).

The national domain

Throughout the history of land ownership, vacant and unowned land has always formed a category of property which has had to be classified separately from that belonging to communities, private individuals and, above all, private and public property. During the German era, the decree of June 15, 1896 (art.1er) classified this land as Reich property in the domain of the German State. A decree dated November 21, 1902 transferred them to the domain of each colony.

On the arrival of the Fancaise administration, this classification was maintained by a decree dated January 12, 1932, organizing the system of state-owned land in Cameroon. This text also incorporated *land which, not being the subject of a regular land title of ownership or enjoyment by application of the provisions of the Civil Code, or of the decrees of July 21, 1932 instituting registration and fixing the mode of establishment of the land rights of the natives in Cameroon, is unexploited or occupied for more than 10 years* (art.1er al.2- dec.12 janvier 1932).

When Cameroon gained independence, the concept of "terres vacantes et sans maître" was abolished. In its place, the decree-law of January 9, 1963 introduced the concept of *national collective patrimony.* Its basis is all land, with the exception of community land, registered or transcribed land, and land in the public and private domain of the State. The State manages the national collective heritage in line with national economic and social development objectives .[17]

Land which, on the date of entry into force of Ordinance 74/2 of July 06 1974 establishing the land tenure system, is not classified in the public or private domain of the State or other legal entities governed by public law, is automatically included in the national domain (Art. 14). Land subject to ownership rights is also not included in the national domain. The outbuildings of the national domain are classified into two categories: residential land, and land used for cultivation, planting, grazing and rangelands whose occupation is evidenced by a clear human hold on the land and conclusive development (Art.15).

---

[17]Jean Marie Nyama, Régime foncier et Domanialité publique au Cameroun, UCAC Press, 2012, p.91

The public domain

Under the terms of article 2 al.1er of ordinance no. 74/2 of July 6, 1974 establishing the domanial regime, public property includes all *movable and immovable assets which, by nature or by destination, are assigned either to the direct use of the public, or to public services.* Public domain assets are inalienable, imprescriptible and unseizable. Subject to certain conditions, they are not subject to private appropriation. Property in the public domain is characterized by its purpose, i.e. its assignment either to public use or to a public service.

An asset can be considered to be in the public interest when its use is freely available to the entire community of a given territory for a purpose of general interest. It makes no difference whether this use is free or paying, collective or private. The use may be natural, as in the case of rivers, or intended by the public authorities, as in the case of roads. Another characteristic of public domain property is that it is used for public service purposes, i.e. for an activity of general interest carried out under the authority of a legal entity governed by public law[18] . This applies to certain public buildings and military commercial ports.

The private domain of the State and other legal entities governed by public law

The private domain of the State, local authorities and public establishments comprises assets of the same nature as those of private individuals. It includes movable and immovable property, as well as real estate rights acquired either for valuable consideration or free of charge. In the first case, acquisition may be amicable [purchase] or forced (expropriation or requisition). In the second case, it may result from the will of the transferors (gifts or bequests), from court decisions (confiscation) or from the law itself (devolution of property without a master and escheated estates). In addition to these two methods of acquiring ownership of property, there are decisions or agreements which, without transferring ownership, confer permanent rights of occupation or use (leases,

---

[18]Joseph Owona, Droit administratif spécial de la République du Cameroun, EDICEF, 1985 p.118

requisitions) on the State, local authorities or public bodies .[19]

Cameroonian law distinguishes between the private property of the State and that of other public corporations. Article 10 of Ordinance 74/2 lists the assets making up the State's private domain. These assets can be divided into two main categories: those which form part of what might be called the State's basic private domain, and those which, because they are vested in the State, form a kind of private domain by incorporation. We will confine ourselves here to enumerating those assets that fall within the basic private domain of the State. The following are thus part of the basic private domain of the State :

1) Movable and immovable property acquired by the State free of charge or against payment in accordance with the rules of ordinary law;
2) Land supporting buildings, constructions, structures and facilities built and maintained by the State;
3) Properties vested in the State under :
- Article 120 of the Treaty of Versailles of June 28, 1919;
- Legislation on war sequestration;
- A deed of classification issued under legislation predating the Ordinance governing the private domain of the State;
- Decommissioning of the public domain ;
- Expropriation in the public interest.
- Rural or urban concessions subject to forfeiture or the right of repossession, as well as the property of associations dissolved for acts of subversion or attacks on the internal or external security of the State.
- Withdrawals decided by the State from the national domain in application of the provisions of article 18 of the land tenure ordinance (Art.10).

The private domain of other persons governed by public law includes :

- Property and real estate rights acquired under private law ;
- Property and real estate rights originating from the State's private domain and transferred to the private domain of the aforementioned persons;

---

[19]J. Magnet, Comptabilité publique, PUF, coll.Thémis, p.262...

- Property and real estate rights acquired under the conditions set out in article 18 of the ordinance governing land tenure (Art.13).

In addition, Cameroon's land tenure law, Law no. 80-21 of July 14, 1980, amending and supplementing certain provisions of Ordinance no. 74-1 of July 6, 1974 establishing the land tenure system, recognizes the allocation of land to foreign investors.

2.2. Awarding concessions to foreign investors

In Cameroon, the concession is defined by article 11 of decree no. 64-10/COR of January 30, 1964, implementing decree-law no. 63 of January 9, 1963, establishing the land and property regime. It consists in granting a concessionaire the right to use a plot of land, with a promise to sell subject to the suspensive condition that the land is developed within a given timeframe. Once this has been achieved, the concessionaire obtains a definitive concession title, which transfers ownership and entitles him/her to a copy of the land title.

Concessions are the ultimate form of land redistribution within the national domain. The concessionaire, invested with a right of occupation that is essentially provisional in origin, can transform the precariousness of his or her possession by developing the concessioned land, of which he or she then becomes the full owner; in a way, it is a potential property subject to a suspensive condition. The concept of a concession varies according to whether it is governed by administrative or civil law.

In administrative terms, a concession to occupy the public domain is "a contract under administrative law conferring on its beneficiary, in return for remuneration, the right to use a more or less extensive part of the public domain". In civil law, it is "a contract by which the owner of a property grants enjoyment of the property in return for annual remuneration, and for a period of at least twenty years, to a lessee who may make improvements of his own choosing and build". In the second case, this is a real estate concession. However, whatever the legal status of the concession, it is first and foremost a voluntary agreement creating reciprocal obligations (art.7 al.2,).

The acquisition of real estate by Diplomatic and Consular Missions accredited to Cameroon may only be authorized on condition of reciprocity. The total surface area that may be transferred may not exceed 10,000m2 for each mission, unless special dispensation is granted by the government. In the event of resale, the State has a pre-emptive right to repurchase the property, taking into account the initial price, the value added and depreciation. Deeds drawn up for this purpose must, on pain of nullity, be submitted for prior approval by the Minister in charge of the Domains.

Allocation of provisional concessions

The provisional concession is granted for development projects in line with the nation's economic, social or cultural options (Art.2). The duration of the provisional concession may not exceed five (5) years. Exceptionally, it may be extended at the reasoned request of the concessionaire (Art.3). Any individual or legal entity wishing to develop an unoccupied or unexploited part of the national domain must submit a request to the domains department in the place where the property is located. This request must be accompanied by, among other things, documents identifying the applicant and the land concerned, as well as a development program outlining the stages involved.

Development is an essential condition of the provisional concession contract. This development of the conceded land must aim to achieve a development project in line with the nation's economic, social and cultural options. As a result, the scope and purpose of the project vary and are assessed in different ways. Concessions of less than 50 hectares are awarded by decree of the Minister in charge of domains, and those of more than 50 hectares by presidential decree. A schedule of conditions sets out the rights and obligations of the concessionaire and the State (Art.7).

Definitive concessions and long leases

The final concession is one of the means of accessing real estate ownership in Cameroon. At the end of the provisional concession period, the Advisory Commission draws up a report on the development of the site, and draws up a report showing the amount of investment made. If the development project is fully completed before the provisional concession expires, the concession-holder may ask the Commission to carry out this assessment. The minutes of this report are used to: a) extend the duration of the provisional concession; b) make the final allocation; c) grant an emphyteutic lease under the conditions set out in article 10 paragraph 3 (Art.9):

The Prefect takes into account the amount of investment made, and may only propose that the land be granted as a permanent concession if it has been developed in accordance with the conditions laid down in the concession deed and any amendments thereto. In the event of partial development of the concession land, the Prefect may request that all or part of the land be granted as a permanent concession. He may only propose long leases for foreigners who have developed a national estate (Art.10).

If not renewed, the lease is terminated on expiry of the initial term. However, it may be renewed, depending on the case, by order of the Minister in charge of estates or by decree, in application of the provisions of article 7. The request for renewal must be made six months before the expiry of the lease; the State may require additional investments at the time of renewal. If the lease renewal application is not approved, or if the lease is terminated, expenses are settled as in the case of leases on the State's private domain (Art.11).

2.2.3. Penalties for infringement of land ownership

In Cameroon, offences against land and state ownership are all punishable by imprisonment, sometimes accompanied by a fine. The custodial sentence can be up to 10 years and the fine up to two million CFA francs, as in the case of falsification of a document attesting to a land right (article 314 paragraph 2-b of the penal code).

Over the years, legislative developments in Cameroon have been characterized by a trend towards ever-increasing severity. For example, infringement of land ownership, formerly punishable by 15 days' to 3 years' imprisonment and/or a fine of 25,000 to 100,000 francs (art. 8 ord. n° 74/1 of July 6, 1974), is now punishable by 2 months' to 3 years' imprisonment and/or a fine of 50,000 to 200,000 francs. To this end, the following are liable to fines and imprisonment: a) those who exploit or remain on a plot of land without the owner's prior authorization; b) public officials convicted of complicity in land transactions likely to encourage the illegal occupation of other people's property.

Conclusion

This entire legal arsenal is reinforced by the administrative protection provided by the Minister of Lands and Land Affairs, and the Minister of Territorial Administration and Decentralization[20] . Under these conditions, the accusations of land grabbing levelled against China in Cameroon are legally unfounded.

---

[20]*Ibid, p.17*

Chapter 3 : China in the market for agro-industrial land concessions in Cameroon

Introduction

Over the past few years, there has been an ever-increasing trend towards large-scale land transfers for agro-industrial purposes. These agricultural transfers encompass all the processes involved in allocating, assigning, transferring or abandoning land to private operators for the purposes of exploitation and economic development. They are carried out either by transfer, alienation, concession or sale. They are not new to the world. They were developed in colonized countries in the form of colonial plantations of rubber, coffee, cocoa or cotton. With the 2008 food crisis, they have increased. The African continent alone accounts for 2/3 of the world's land acquired in this way. Cameroon is no exception. Today, it is one of the countries with the highest concentration of agro-industrial land sales. There are several reasons why land in this country is so attractive to foreign agribusinesses, including favorable agro-ecological conditions and relatively cheap land prices. Many companies have already secured land here, and others are still negotiating. Although official statistics on the areas actually allocated are not yet available, between 2014 and 2015 the Ministry of Domains, Cadastre and Land Affairs counted more than twenty companies allocated or in the process of acquiring land concessions. Given the objectives and ambitions of the Cameroonian government's economic growth policies and strategies, it is not out of the question that requests and allocations will intensify and multiply over the next ten years.

A number of features underpin the phenomenon of land allocation in Cameroon, notably the diversification of investors. Whereas in the past it was classic agribusiness companies that became involved during privatizations, there is now a growing interest in land concessions by large multinationals, foreign national companies and large local enterprises. The size of concessions continues to grow. Traditional export crops [cocoa, coffee, cotton, etc.] are being joined by new products such as corn, rubber, jatropha and oil palm .[21]

According to FAO statistics, Cameroon has around 6.2 million hectares of arable land, of which 1.3 million hectares, or just over 20%, are cultivated. This potential, together with Cameroon's agro-ecological diversity, easy access to the sea and immense irrigation potential - 240,000 hectares of potentially irrigable land for agriculture, of which only 33,000 are currently irrigated - make the country particularly attractive for investment in the agricultural sector. The increasing demand for arable land, not only at global level but also at national and local level, is a trend that has been observed in recent years, and one that Cameroon has not escaped. This trend has taken two main forms: a) *Increased investment in agricultural land.* This is taking place both in large-scale agro-industries and in medium-sized farms (from a few dozen to a few hundred hectares, mainly controlled by local or national elites);b) *The pressure exerted on agricultural land by non-agricultural activities.* Major infrastructure projects (dams, pipelines, railroads, deep-water ports, etc.), logging and mining concessions and protected areas all encroach on land used by communities (or that could be used in the more or less near future). Their rapid development, and above all their concomitant occurrence, is exacerbating the land shortage in rural areas. These developments are taking place at a time when population growth is likely to lead to a significant increase in demand for arable land in rural communities .[22]

The perception of the abundance and availability of land in Cameroon is fuelling economic competition between China and the major powers in the country, to the extent that some media and NGOs are overly enthusiastic about the idea of a Chinese land grab in Cameroon. But what is the truth? Before answering that question, let's take a look at the overall situation of the land concessions market in Cameroon.

3.1 Overview of agro-industrial land acquisitions in Cameroon

Most researchers working on the question of large-scale land acquisitions in Cameroon,

---

[21]Samuel Nguiffo and Michelle Sonkoue Watio "Investments in the agro-industrial sector in Cameroon. Large-scale land acquisition since 2005", pp.2-3
[22]Ibid, p.41

including ourselves, agree that it is difficult to characterize agricultural investments by their size. Most transactions do not always comply with the requirements of national land law, particularly those falling within the exclusive remit of the Ministry of Cadastre, Domains and Land Affairs. Hence the great difficulty of systematically establishing an exhaustive list of all large-scale land acquisitions in Cameroon.

Table 1. Agro-industries present in Cameroon before 2005					
Companies/agricultural jet	Shareholders	Location	Area allocated by agreement with the State [hectares]	Speculation	Existence of a long lease
CDC	Parastatal	Various in the South-West	102 000	Oil palm, rubber, banana	Yes
Pamol	Private 90 Condition 10		41 000	Oil palm	Concession
Socapalm	Bolloré 70	Various in the Littoral and South	58 000	Oil palm	Yes
hevecam Golden Milenium Group [GMG]	Private [GMG] 90	Ocean Department [South]	41 000	Hevea	N.c [not known]
Upper Noun Valley Development Authority	N.c	Ndop	136 700	Rice	N.c

[UNVDA]					
Cameroon Sugar Company [SOSUCAM]	N.c	Mbandjock Nkoteng	12 000	Sugar cane + sugar processing	N.c
Mount Mbappit rural development project [MINADER]	N.c	Noun	1200	Rit + market garden produce	N.c
SOCAPALM [formerly Swiss farm]	N.c	Edéa	3793	Oil palm + palm oil processing	N.c
SOCAPALM [formerly SAFACAM]	State and private Cameroonian and foreign	Dizangué	4870	Rubber + oil palm	N.c
Village plantations, Sanaga Maritime	N.c	Sanaga Maritime		Rubber + oil palm	N.c
Ndawara Tea Estate	Private	Northwest	N.c	Tea	N.c

Source: Samuel Nguiffo and Michelle Sonkoue Watio, 2015, pp.14-15

Table 2. Land acquisitions in Cameroon in LAND MATRIX

Target Country	Primary Investor	Secondara y investor	Secondar y Investor country	Intention of investmen t	Negotiatio n Status	Impleme ntation Status	Intended Size (ha)	Contra cted Size (ha)
Cameroo n	Palm Resources Cameroon Limited	Biopalm Energy Limited	Singapore	Agricultur e	[2011] Included (Contact signed)	[2014] Project not started	200 000	3 348
Cameroo n	South- Cameroon Hevea S.A	GMG Global Ltd SPPH	Singapore France	Agricultur e	[2013] Included (Contract signed)	[2013] startup Phase (no productio n)	65 000	45 000
Cameroo n	Plantations From top P^{enja} (PHP)	Fruit company	France	Agricultur e	[2014] Included (Contract signed	[2014] Project not started	800	800
Cameroo n	Justin Sugar Mills	Justin Sugar Ltd	United Kingdom of Great Bitain and Northern Ireland	Agricultur e	[2013] Included (Contract signed)	[2014] Startup Phase (no productio n)	54 632	54 632
Cameroo n	Sino-Cam Iko Ltd	Shaanxi Land Reclamatio n General Corporation	China	Agricultur e	[2006] Conducted (Contract signed)	[2010] Startup phase (no productio n)	10 120	10 120

Camerou n	SG Sustainabl e Oils Cameroon,	Heraklès Capital	United States of America	Agricultur e	[2013] Conducted (Contract signed)	[2011] Startup phase	100 000	19 843
	Ltd. (SGSOC)					(no productio n)		

Source: Land Matrix

Comparing these two tables, it's obvious that they contain more or less different data. The major difficulty here is that it is very difficult to collect reliable data on large-scale agricultural investments in Cameroon. This is due to the absence of a government database centralizing all transactions in this field. Similarly, many announcements are made in complete disregard of reality. Certain announced investments sometimes take place in different areas and on different surfaces, without the initial announcements being modified in the press and other publicity documents.

The only certainty is that these land acquisitions are increasing, and the players also fluctuate over the years, including the crops envisaged. Between 2005 and 2013, for example, nearly 40 agribusinesses applied for and obtained some form of agreement granting them rights over land in the national domain.[23] Most of the investors who have chosen Cameroon since 2005 are foreign companies, sometimes in partnership with nationals (individuals or companies), enabling some to conclude agreements directly with local communities, bypassing the relevant national land legislation, or with the State, in the form of lease contracts, provisional concessions or tacit agreements.

In terms of the evolution of these land transactions, between 2009 and 2012, there was an upsurge in the number of requests for arable land and a significant increase in the surface areas requested. In the end, the number of land transactions concluded (provisionally or

---

[23]Samuel Nguiffo and Michelle Sonkoue Watio "Investment in the agro-industrial sector in Cameroon. Large-scale land acquisitions since 2005", IIED and CED, 2015, p.28

definitively) was much lower than the number of transactions announced. Thus, for the period under review, a total of 11 transactions were recorded in the category of land under provisional or definitive concessions (1 provisional and 10 definitive), i.e. a third of the 33 transactions announced in all. However, in some concessions, activities (such as milestones, demarcation, felling, nurseries and production) had begun while the negotiation process was still underway. All in all, more concessions are currently in operation than have final or provisional concessions[24] . The areas applied for over the past ten years are increasingly larger than those prior to or since 2005.

---

[24]Ibid, p.29

Table 3. Duration of land concession contracts allocated to the agro-industrial sector in Cameroon	
Company concerned	**Lease term**
SGSOC [SG Sustainable Oils Cameroon]	99 years in the establishment agreement, and 3 years of provisional concession in the presidential decrees of November 25, 2013
Sino Cam Iko Agriculture Development	20 years, possibly renewable by mutual agreement, from the date the sites are made available[21]
HEVECAM	99 YEARS
SOSUCAM	90 YEARS
Pamol Plantations PLC	90 years old
Socapalm	60ans
CDC	60
West End Farms	50 years
Tchassen Holding	30 years
PHP [Société des plantations du Haut Penja].	25 years old
Sagex	10 years

| Kawtal Demi | 10 years |
| Biopalm Energy Ltd | 3 years [provisional concession]. |

Source: Samuel Nguiffo and Michelle Sonkoue Wation

Alongside these requests from agribusinesses, there are also individual requests for a total of 800,000 hectares. The most publicized allocations are often for large areas. Many people lose sight of the magnitude of smaller acquisitions by the elite.

National elites and companies belonging to them are taking control of an ever-increasing area of land. These acquisitions are generally made in the communal areas of the buyer's community of origin, thus mobilizing two contradictory logics to satisfy his personal interests: the community logic, which gives him access to the communal areas of the village land, and the state logic, which gives him the right to appropriate it privately.

It should be pointed out that it is even more difficult to obtain information on areas controlled by private companies owned by Cameroonians. Whether foreign, mixed or national, these investors all want to produce crops with good if not high profitability on the local and international markets. This no doubt explains why MINADER authorities decided, at the end of 2013, to redirect some of these investors towards other crops.

3.2. Chinese land "grabs" in Cameroon: wishful thinking or reality?

As in many African countries, Cameroon, through the voice of its highest state authority, has adopted an economic policy focused on attracting foreign direct investment as the engine of growth for Cameroon's emergence by 2035. In the land sector, this objective is reflected in a presidential instruction aimed at encouraging large-scale agricultural investment, referred to as second-generation agriculture, and in the need to adopt reforms to facilitate access to land for investors in the agricultural sector. Since this landmark announcement, requests from international and national investors have been pouring into the three ministries (MINADER, MINFOF and MINDCAF) in charge of the various aspects of land ownership.

In this race for land, where major economic powers such as the European Union, the United States and China are jostling for position, land access strategies differ from one player to another. While some go through legal channels by concluding agreements with the state, others negotiate directly with communities through nationals who have previously acquired the land, often for a symbolic franc. In this strategic confrontation, China is accused of land grabbing in Cameroon. What is the fundamental situation? This is the main concern of this book. Before plunging into any analysis, we need to present a general overview of Chinese land acquisitions and solicitations in Cameroon.

Table 4: Areas requested or acquired by China in Cameroon						
Company	Area Solicited	Areas obtained	Developed areas	Existence of a long lease	Location	Agricultural project	Extensio n Planned
HEVECA M - GMG	41 000	41 000	22 000	YES	Nieté Département De l'Ocean Sud	Hevea	N.c
South-Cameroon Hevea	45 200	45 200	N.c	YES	Meyo messala Meyomessi Djoum Southern region Dja et Lobo	Palm oil, Hevea	N.c
Group Chinese	4000	N.c	N.c	N.c	N.c	Rice, breeding	N.c
Sino Cam Iko Agriculture	10.0004ha	6000ha	100ha	YES	Nanga-Eboko [5000 ha] Ndjoré [1000ha] Santchou [4000ha] Noun [2 ha] Limbe [2 ha]	Rice Cassava Soybeans Maize Livestock	Yes

Source: MINADER, MINFOF, MINDCAF surveys

This table shows the existence of a lease between these Chinese agribusinesses and the State of Cameroon. This undoubtedly implies the existence of a contract, and therefore of a legal approach in terms of land access strategy.

GMG (a subsidiary of Sinochem International) acquired the former Cameroonian company HEVECAM in 1996, as part of the state privatization process. As for Sud-Cameroun Hévéa, 80% owned by the Singapore-based Sinochem Intenational Group, it was following an audience granted by the Secretary General of the Presidency of the Republic of Cameroon, Mr. Ferdinand Ngoh Ngoh, to Mr. Liu Deshu on May 24, 2011, that a $410 million agreement was signed in December 2011 between this firm and the Cameroon government, for the development of 45.200 hectares of oil palm and rubber plantations in Meyomessale, Meyomessi and Djoum in the southern region. As for Sino Cam Iko Agriculture, its presence in Cameroon's agricultural land sector follows the signing of a Memorandum of Understanding on January 13, 2006, and a Framework Agreement on March 24, 2010, between the Cameroon government, *Integrate-Industry-Commerce Corporation Of Shaanxi reclamation & States Farms* and the *China Development Bank.*

Conclusion

The many accusations of Chinese land grabbing in Cameroon are a myth. The table on Chinese land solicitations and acquisitions in Cameroon is ample proof of this. Despite these few elements, and in view of the complexity of this phenomenon on the national scale of Cameroon, far from having the ambition of being exhaustive, this demonstration has a more modest vocation, and is content simply to indicate trends, on the basis of information collected during our various readings and field surveys. In a way, this book is a contribution to reflection on the question of land acquisitions in the Cameroonian context, and as such, attempts to provide an overview of the situation of agro-industrial land transactions in this country, particularly as regards China, while indicating the applicable law in the matter.

Chapter 4: Understanding the issues surrounding accusations of Chinese land grabs in sub-Saharan Africa

Introduction

China's eclectic presence in the economies of sub-Saharan Africa in recent years seems to have fundamentally irritated the traditional powers, who have long regarded Africa as a preserve in the pay of the major global financial institutions, whose vision of Africa's development seems at odds with what Africa wants of itself. By way of example, the dark period of structural adjustment that plagued Africa at macro-economic level is quite illustrative of the diktats imposed on Africa. At the micro level in Cameroon, the State was forced to privatize a large number of parastatal companies, most of which were taken over by large Western groups.

Africa is also plagued by poverty, pandemics of all kinds, recurrent civil and tribal wars, political instability and the desertion of certain traditional players. Faced with this situation, China presents itself as an alternative for an Africa aspiring to emerge from the depths of underdevelopment. China seems to have understood better than Western players that Africa has deficits in all areas of its economy, especially in terms of its infrastructure, which for the most part dates back to the colonial period. This seems to justify the development model proposed by China. A model that represents a complete break with Western models, which are conditioned by the practice of democracy, transparency and, above all, good governance.

Unlike Western players and, by extension, International Financial Institutions (IFIs), China does not impose any preconditions such as democracy or respect for human rights (Courmont & Lewis, 2007; Carmody & Owusu, 2007). In the eyes of African leaders, China represents the hope of another world that gives priority to bread over the right to vote (Ndubisi Obiorah, 2011).

The virulence of Western criticism of China is nevertheless astonishing, especially given that for a long time, at least until the early 2000s, Africa did not attract Western investors as it does today. Indeed, both the USA and the European Union had neglected it after the Cold War. Paradoxically, the EU had maintained its relations with North Africa, whereas it took a very long time for it to deign to define a policy towards sub-Saharan Africa.

For over fifteen years, the EU has been heavily involved in North Africa through three initiatives: the Barcelona Process (1995), the European Neighborhood Policy (2004) and the Union for the Mediterranean (2008). It was not until December 2005 that the EU presented its *EU Strategy for Africa*, based on three principles: 1) equality, based on reciprocal recognition and respect for institutions and the definition of collective and mutual interests; 2) partnership, i.e. developing links based on commercial and political cooperation; 3) ownership, implying that development strategies and policies should be specific to the countries concerned and not imposed from outside. (Struye de Swieland, 2010). In October 2008, the EU proposed a document entitled "The EU, Africa and China: towards trilateral dialogue and cooperation", a series of trilateral initiatives based on three principles - pragmatism and progressiveness, a shared approach, aid effectiveness - and four concrete objectives to be achieved. On the ground, these declarations and initiatives have remained dead letters, with China preferring to follow its own policy .[25]

The criticism that abounds about CHINAFRIQUE can probably be partly explained by China's rejection of the idea of a China-Europe partnership on Africa. This Western critical stance towards China could also be explained by the fact that :

- China refuses to adhere to certain standards adopted by DAC countries,
- It continues to maintain relations with states that have been ostracized by the international community,
- It refuses to adhere to the target of devoting 0.7% of its gross national income to official development assistance,
- Its method of accounting for aid differs from that adopted by DAC countries,

- It continues to maintain a blockade on several other non-rare metals in addition to the "rare earths"[26] for which it was forced by the World Trade Organization at

---

[25]*Ibid,* p.112.

[26]Rare earths, also known as "lanthanides" (Cerium, Neodymium, Europium, Dysprosium, Terbium, Yttrium...), play a key role in the electronics industry (screens, hard disks) and in green technologies (wind

the end of 2011 to lift its export restrictions via quotas, following a complaint lodged by the United States, the European Union and Japan.

4.1. The debate on the Chinese presence in sub-Saharan African agricultural land

China's presence in sub-Saharan Africa's agricultural land is the subject of much controversy. Many accuse China of engaging in large-scale land grabs, whereas data from the Land Matrix (2013) invalidates this perception, showing, on the contrary, that China's public or private land acquisitions amount to just 290,000 hectares, 15 times less than the United Arab Emirates and more than 6 times less than the UK[27] . Nor is Africa a priority continent for China, which acquires almost twice as much land in South America and Southeast Asia .[28]

China is also wrongly regarded as a major financier of agriculture in sub-Saharan Africa. For despite the fact that its funding has been steadily increasing since 2008, the amount of its aid remains low, compared to that of OECD countries, which amounted to around $130 million between 2009 and 2012, against $3 billion in commitments by DAC bilateral and multilateral donors to agriculture and rural development in 2012 alone.

4.2. Clarification of accusations of land grabs Chinese in Cameroon

Cameroon's accusations of land grabbing by China are pure invention and misinformation. Cameroon has a whole legal arsenal for land management, notably the colonial decree of July 21, 1932 instituting the land registration system in Cameroon, followed by ordinances no. 74/1 and 74/2 of July 06, 1974 establishing the land tenure

---

turbines, hybrid cars). With 97% of global production dominated by China, rare earths have become a major issue for high-tech industries and governments alike. The Chinese understood the strategic value of these metals as early as the 1980s. At the time, other countries relied on free trade to guarantee easy access to these resources, and therefore did not develop their production. China, on the other hand, did so, with little regard for environmental constraints, enabling it to replace the United States as the world's leading producer.
Foreign countries see China's attitude as a threat. To get out of the crisis, they are betting on high technology, which needs rare earths. China has more or less the same goal: to reorient its economy towards more added value. Rare earths are a major component of its power strategy. In particular, it has used them against Japan, with whom it is in conflict over territorial issues in the China Sea.

[27]Jean-Jacques Gabas and Xiaoyang Tang, *Ibid.*
[28]*Ibid.*

and property regime, together with their implementing decrees.

Cameroon's land tenure system - and this is the major concern of this book - provides for the transfer of land to foreign investors through Law n°80-21 of July 14, 1980, amending and supplementing certain provisions of Ordinance n°74-1 of July 6, 1974 establishing the land tenure system.

Cameroon's land tenure system also protects land ownership through the existence of penalties for infringement of land ownership. There are also mechanisms for resolving land disputes, notably jurisdictional and administrative measures. However, as the law is the product of men, it would be illusory to think that Cameroon's land law escapes the trivial reality common to almost all societies, namely the gap between texts and their application. Hence the many dysfunctions

The willingness to pervert, truncate or deliberately distort information, which seems to characterize certain Western media today, nevertheless raises a problem of ethics in the journalistic profession. This raises the question of the journalist's professional duties. In France, the Charte des journalistes professionnels, whose spirit remains the same, was adopted by the SNJ in July 1918 and amended in 1938. This Charter stipulates that:

"a journalist, worthy of the name, takes responsibility for all his or her writings, even if anonymous; considers slander, accusations without proof, alteration of documents, distortion of facts and lies to be the most serious professional faults; recognizes only the jurisdiction of his or her peers, who are sovereign in matters of professional honor, accepts only assignments that are compatible with professional dignity, refrains from invoking an imaginary title or capacity, from using unfair means to obtain information or surprise anyone's good faith, does not receive money from a public service or private enterprise or his or her status as a journalist, does not sign his name to commercial or financial advertising articles, does not commit plagiarism, does not quote colleagues from whom he reproduces any text, does not solicit the position of a colleague, nor provoke his dismissal by offering to work under inferior conditions, maintains professional secrecy, does not use the freedom of the press for self-interested purposes, claims the freedom to publish his information honestly, holds scrupulousness and concern

for justice as primary rules, does not confuse his role with that of a policeman"[29] .

Unfortunately, today's clichés about CHINAFRIQUE suggest that journalism is governed solely by the rules of editorial marketing. Of course, it's important to attract interest, otherwise the newspaper will die. But journalists must also have a social function, and ask themselves what the purpose of their article is. In our opinion, journalists should describe reality as it exists, without blush or taboos - that's their job. Journalists are not judges, whose function is to hand down sentences. Nor is the journalist's job to do harm to others or to praise the "powerful". Faced with information and media flows overflowing on all sides, journalists can remain landmarks if they know how to select the relevant information and prioritize the different pieces of information. Otherwise, as sociologist Jean-Marie Charon puts it[30] , the situation could result in a "great misunderstanding" between journalists and their audience. Some would feel, for example, that they are doing their job to the best of their ability, while others would feel that journalists are not meeting their expectations.

That's why we're delighted to have been out in the field, and to have been able to provide a few elements of clarification on China's presence in Cameroonian agricultural land tenure, based on the methodological advice of Emile Durkheim, who, in *Les Règles de la méthode sociologique (The Rules of Sociological Method), sets out* a three-stage methodological approach: definition of the phenomenon, refutation of previous interpretations, and a properly sociological explanation of the phenomenon under consideration. Durkheim's methodological approach is an invitation to rid oneself of preconceptions, presuppositions and, ultimately, received ideas, in order to submit rigorously to the discipline of methical doubt.

In keeping with Durkheim's methodological approach, we have opted for the comprehensive approach. Comprehensive sociology is characterized by a triangulation between theory, field and researcher. It calls into question the idea of the researcher's exteriority in relation to his or her object of study. Today, this approach prevents us from sounding like the journalists whose methods we decry. Indeed, in the tradition of Pierre

---

[29]Francois Jost, 50 fiches pour comprendre les médias, Paris, Bréal, 2009, p.26
ss[27] Quoted by Francois Jost, Op.cit.

Bourdieu, sociologists have rightly developed a critique of the "journalistic field". The main thrust of this critique is that newspapers are in the business of making money, that the commercial has taken precedence over the editorial and that, while not all journalists are sell-outs, many of them accept to work for advertisers and are subject to pressure from all quarters. We largely agree.

However, even if China and other foreign players in the land concessions market in Cameroon can be exonerated from accusations of land grabbing, the Cameroonian state, which we deem partly responsible for spreading rumors of Chinese land grabs in Cameroon, cannot be exempted from all blame. This is because foreign, and in particular Chinese, land acquisitions are taking place in Cameroon in a social context marked by the impact of modern land laws on the land rights of local and indigenous communities.

This extraordinary situation has its origins, both politically and legally, in the persistent application of certain colonial laws. During the era of colonialism, it was convenient to deny "natives" ownership of the land on which they and their ancestors had always lived, in order to appropriate the majority of resources. Land was proclaimed state property, and traditional owners were designated as mere occupants and tolerated users of the land. Virtually the whole of sub-Saharan Africa was affected [the situation of indigenous peoples is almost identical]. From the outset, collective properties by their very nature, such as forests and pastures, were the most vulnerable to expropriation. Even when land was actively occupied and used, the interest of the state [the government, to be precise] took precedence. Over time, these measures were legitimized by the interests of the emerging political and economic elites who monopolized the land.

In the case of Cameroon, the government went one step further. It took the opportunity, via its 1974 land laws, to remove the opportunities created by colonial land laws to enable rural communities to register their own estates, thereby offering them a degree of protection[31] . Aware that this situation is likely to deteriorate further in the long term, we suggest that the Cameroonian state, from the point of view of efficient management of Cameroon's vast land heritage, develop an inclusive policy that integrates the cohabitation

---

[31]Liz Alden Wily, "Whose land is it? The status of customary land ownership in Cameroon", Ed Fenton, 2011, p.13

of written and customary law. For even if Cameroon offers the spectacle of a veritable ethnic mosaic, the size of which varies from one region to another, from one people to another, and even if it has more than 230 ethnic groups, at least if we consider the languages spoken, and even if in such a context of cultural plurality, modes of existence and worldviews are not necessarily identical, let alone automatically compatible, there are nevertheless certain general principles, certain common elements which, in the field of traditional land law, allow us to speak of a common conception of traditional land ownership.

While the primary aim of this book is to shed light on the accusations of Chinese land grabbing in Cameroon, the second objective is to urge Cameroon's public authorities to adopt an inclusive approach to land management. In this respect, the Cameroonian state could, for example :

- Attempting to return to the historical sources of land law, by once again developing a system of customary land ownership built around traditional institutions.
- To delimit community territories within the framework of a collective property regime materialized by a land title established in the name of the community. The community land title would thus have the same attributes as other land ownership certification documents.

This would leave us with the specific question of indigenous forest communities, who are not familiar with private ownership of land or natural resources, and whose relationship to land and forest could hardly be covered by the common resource regime, in the sense that they do not envisage the exclusion of non-natives from forest use (Wily, 2008).

Conclusion

While the reports of Chinese land grabs in Cameroon carried by the international media and NGOs are, in our view, problematic, so too is the attitude of the Cameroonian state to the social context in which foreign land acquisitions prevail: legal pluralism, collision between modern and customary law, elitist land grabs, lack of systematic public information on agro-industrial land concession activities, slowness of the State in taking decisions to sanction the marginal behavior of certain foreign agro-industries ,marginalization of indigenous populations from the process of allocating and managing land concessions, are all factors that may lead the general public and certain local people to give credence to claims of Chinese land grabbing in Cameroon.

General conclusion :

Lack of tools for systematically informing the general public about agro-industrial land concession mechanisms

Cameroon's legal system is made up of rules classified into two categories: a) rules of internal origin, which emanate from Cameroonian institutions or authorities; b) rules of international origin, which are provisions governing access to information that emanate from international bodies and are applicable as mandatory rules on Cameroonian territory. With regard to the rules of internal origin to which we shall confine ourselves, the rules and institutions relating to transparency in the process of allocating and managing land concessions can be found in the Constitution, in particular Law no. 90/053 of December 19, 1990 governing associations, Ordinance 74-1 of July 6, 1974 laying down the land tenure system, Decree 76-166 of April 27, 1976 laying down the procedures for managing the national domain, and Decree 76-165 of April 27, 1976 laying down the conditions for obtaining a land title. This legal framework is supplemented by Instruction n° 000006/Y.18/MINDAF/D300 of December 29, 2005 relating to the functioning of the Consultative Commission, particularly with regard to aspects concerning formalities prior to the allocation of land for a project.

However, an analysis of these texts reveals that Cameroon's legal and institutional framework for access to information in the process of allocating and managing land concessions has two main features: 1) it is unfavorable to direct access to information; 2) it favors indirect access to information. Apart from the publication of allocation notices at the prefecture and sub-prefecture where the land is located, the publication of an order or decree is an operation that escapes communities and civil society organizations, who only incidentally consult the Journal Officiel, which is not sufficiently accessible. This only provides information on the outcome of the allocation operation.

However, three instruments give the public direct access to information on land transactions. These are: a) the land policy document; b) the land use plan and the land registration instrument; c) the land register.

The land policy document is the document by which the State indicates the major orientations it intends to give to the allocation of land within its territory. It is the first instrument of transparency, enabling citizens, investors and other interested parties to know what the legislator's intentions are, and whether or not concessions have priority in the choices made.

The land-use plan is the document by which the authorities divide up the land of the territory and indicate the destination of each space or block of space. It enables potential investors and the public to know what land is available for concessions. It allows those with a claim to the land to make themselves known.

The Cameroonian government has launched studies to produce these two documents, which are still in the process of being drawn up, and are therefore of no help to anyone seeking information on current or future land transactions in Cameroon.

The *land register*. In Cameroon, this document only records transactions on registered land, which is not the case for concessions. It does record definitive concessions, but only with a view to transforming them into title deeds. This means that it provides no information on the process of allocating and managing land concessions.

What's more, the contracting parties are under no obligation to publish the information. In fact, there is no text in Cameroonian law requiring the parties to a transfer contract to make the content of their agreement available to third parties. In this respect, Cameroon strictly adheres to the principle of the relative effect of contracts, which means that the contract is the law of the parties. This is laid down in article 1165 of the Civil Code, which states that: "Agreements have effect only between the contracting parties, they do not harm third parties, and they benefit them only in the cases provided for in article 1121"[32] . Since contracts do not harm or benefit third parties, they need not be brought to their attention. The parties may even, by means of an express clause, impose on themselves an obligation not to disseminate information relating to their agreement. They therefore have the right to conduct negotiations in a secretive manner, and not to disclose the content of

---

[32]Article 1121 of the French Civil Code (·) deals with stipulations for third parties, where a contract is intended to produce effects in favour of a person who was not a party to the contract.

their agreement to third parties. However, the concession contract,

although concluded by the State, benefits, or could indirectly harm, third parties, in particular neighboring populations, in social and environmental terms. It is certainly to put this principle into perspective that international guidelines and principles in this field recommend that States consider third parties in such transactions .[33]

In concrete terms, in Cameroon, the evolution of the level of knowledge that communities have about a company acquiring land is linear according to the following schematization: name of the company, activities carried out, nationality of the managers, destination of the products, date of establishment, existence or not of an extension project, area conceded, and duration of the lease contract. The information most readily available concerns the name of the company and the nature of its activities. On the other hand, two pieces of information crucial to a balanced partnership between communities and the company are virtually unavailable at local level: the surface area granted to the company and the duration of its contract.

---

[33]Pierre Etienne Kenfack, Le cadre légal et institutionnel de l'accès à l'information dans le processus d'attribution et de gestion des concessions foncières agro-industrielles au Cameroun, RELUFA, 2015, PP.26-27

Bibliography

General works

Autret (Florence.), *Les manipulateurs. Le pouvoir des lobbys,* Paris, Denoel, 2003, 240p

Aron (Raymond.), *Les étapes de la pensée sociologique,* Paris, Gallimard, 2010, 662p.

Ayissi (Lucien.), *Corruption et gouvernance,* Yaoundé, Presses Universitaires, 2003, 187p.

Badie (Bertrand.), *La diplomatie de l'intrus : l'entrée des sociétés dans l'arène internationale,* Paris, Fayard, 2008, 283p.

Badie (Bertrand.), *La diplomatie de connivence : les dérives oligarchiques du système,* Paris, La Découverte, 2011, 270p.

Bacqué (Marie-Hélène.), and Sintomer (Yves.), *La démocratie participative : Histoire et généalogie,* Paris, La Découverte, 2011, 288p.

Baillet (Jerôme.), *l'exclusion : définitions et mécanismes,* Paris, Hachette, 1996,209p.

Breton (Philippe.), *La parole manipulée,* Paris, La Découverte, 2000, 225p. Caillé (Allain.), *La quête de reconnaissance : nouveau phénomène social,* Paris, La Découverte, 2007, 304p.

Coste (thierry.), *Le vrai pouvoir d'un lobby. Les politiques sous influence,* Paris, Bourin, 2006, p.317.

Colonomos (Ariel.), *La morale dans les relations internationales,* Paris, Odile Jacob, 2005, 368p.

Dérathé (Robert.), *Jean-Jacques Rousseau et la science politique de son temps*, Paris, Vrin, 1995, 492p.

Flahault (François.), *Où est passé le bien commun?,* Paris, Mille et une nuit, 2011, 256p.

Goguel d'Allonsdans (Alban.), *L'exclusion sociale : les métamorphoses d'un concept (1960-2000),* Paris, L'Harmattan, 2003, 167p.

Gueyenot (Claude.), *L'insertion : discours, politique et pratique,* Paris, L'Harmattan, 1982,221p.

Howard (Becker.), *Comment parler de la société,* Paris, La Découverte, 2009, 316p.

Jost (Francois.), *50 fiches pour comprendre les médias,* Paris, Bréal, 2009,158p

Kubler (Daniel.), Maillard (Jacques.), *Analyser les politiques publiques,* Grenoble, Presses Universitaires, 2009, 224p.

Lemieux (Vincent.)*, L'Etude des politiques publiques : les acteurs et leur pouvoir, 2eme edition,* Laval, Presses de l'Université, 2002, 195p.

Michel (Richard.),*Les doctrines du pouvoir politique,* Paris, Collection Synthèse, 1986, 189p.

Monondzana (Hubert.), "La mondialisation: gouvernance et transparence", Yaoundé, 2004, 106p.

Orsena (Erik.), *Un monde de ressources rares,* Paris, Perrin- Descartes et Cie, 2007, 209p.

Rousseau (Jean-Jacques.), *Du Contrat social,* Paris, Garnier-Flammarion, 2001, 286p.

Santander (Sébastien.), *Emerging powers: a challenge for Europe?* Paris, Ellipses, 2012, 382p.

Sassen (Saskia.), *Expulsions: Brutality and Complexity in the global economy*, Belknap Press, 2014, 304p.

Sassen (Saskia.), *The Global City: New York, London, and Tokyo, second edition, Princeton, 2001.*

Stiglitz (Joseph. E.), *La grande désillusion,* Paris, Fayard, 2003,407p.

Ziegler (Jean.), *La haine de l'occident,* Paris, Albin Michel, 2008,304p.

Special works

Bachelet (michel.), *Systèmes fonciers et réformes agraires en Afrique noire,* Paris, L.G.D.J, 1968.

Chabas (Jean.), *La propriété foncière en Afrique noire,* Paris, 1957

Le Roy (Etienne.), *La terre de l'autre : une anthropologie des régimes d'appropriations foncières,* Editions LGDJ, Collection droit et société, Paris, 2011, 448p.

Le Roy (Etienne.), and Bouju (Jacky.), *Prendre en compte les enjeux fonciers dans une démarche d'aménagement,* Paris, GRET, 2004, 128p

Vanderlinden (Jacques.), et Le Roy (Etienne.), *La terre et l'homme - Espace et Ressources convoités, Entre le Local et le Global,Paris,* Karthala,2013, 324p.

Specific articles and reports

FAO document archive, "Land tenure and rural development: What is land tenure?", 2013

Charvet (Jean-Paul.), "Land grabbing or accaparement des terres agricoles", Encyclopaedia universalis [online], accessed February 25, 2014- URL : http//www.universalis-edu.com/encyclpédie/landgrabbing or accaparement-de-terres-agricoles

Chouquer (Gérard.), "Understanding massive land acquisitions in the world today",2012.

Chouquer (Gérard.), "Glossaire des acquisitions massives des terres (ou land grabbing), 2012.

Chouquer (Gérard.), "Aspects et particularité de la domanialité en Afrique de l'Ouest", DES FICHES PEDAGOGIQUE, pour comprendre, se poserde bonnes questions et agir sur le foncier en Afrique de l'Ouest, www.agter.org/bdf/fr/corpus chemin/fiche-cemin-45.html.

Transnational Insitute Programme "Land grabbing", *Agrarian Justice,* February 2013
Ward Anseeuw, Liz Alden Wily, Lorenzo Cotula, and Michael Taylor, *"Land rights and the land rush",* iied, cirad, ILC, 2012.

Rights and Resources Initiative. Annual report, *"Landowners or landless peasants: What choice will developing countries make?",* 2012-2013.

Books, articles and reports on land concessions in sub-Saharan Africa

Works

Cotula (Lorenzo.), Vermeulen (Sonja.), *Land grab or development opportunity? Agricultural investment and international land deals in Africa,* IIED/FAO/IFAD, London &Rome, 2009, 120p.

Reports and articles

Bruno (Pierre.), *"Appropriations foncières: après l'affaire Dewoo, que se passe-t-il à Madagascar?",* June 2011.
Cotula *(Lorenzo.),Acquisitions foncières en Afrique que disent les contrats fonciers,* IIDE, 2011, 66p.
Frierich Ebert Stiftung Foundation, *"Pladoyer pour un réforme du régime*

juridique des cessions foncières à grande échelle en Afrique Centrale", Framework document, Presse universitaire d'Afrique, 2013.

Gabas (Jean-Jacques.), and Xiaoyang (Tang.)," *Coopération agricole chinoise en Afrique subsaharienne : dépasser les idées reçues"*, Perspectives n°26 February 2014.

Lasserve (Alain-Durand.), & Le Roy (Etienne.), *"La situation foncière en Afrique à l'horizon 2050"*, January 2012.

NGUIFFO (Samuel.), KENFACK (Pierre-Etienne.), MBALLA(Nadine.), *"Les droits foncier et les peuples de forêts d'Afrique: Perspectives historiques, juridiques et anthropologiques"*, January 2009.

Schonecke (Wolfgang.), *"The lands of Africa. The new Eldorado"*, 2009.

Books, articles and reports on land concessions in Cameroon

Elong (Joseph-Gabriel.), *L'élite urbaine dans le paysage agricole africain : exemple camerounais et sénégalais,* Paris, L'harmattan, 2004.

ACDIC, *"La cession des terres aux entreprises étrangères: Quels enjeux pour Le développement au Cameroun? What opportunities, what risks?"*, 2009.

African Development Bank &African Development Fund, "Cameroun: Etude diagnostique pour la modernisation des secteurs du cadastre et des domaines", 2009

Madingou (Edouard.), *"Les conflits liés à la gestion décentralisée des ressources forestières au Cameroun : Etat des lieux et perspectives"*,

2011.

GRAIN, *"Demasquer les accaparements de terres camerounaises par une entreprise Chinoise"*, October, 2010,

Greenpeace, *"Herakles Farms in Cameroon: a counter-example for palm oil"*, February 2013

Hoyle (David.), Levang (Patrice.), "Oil palm development in Cameroon", April 2012

Ndjogui (Thomas Eric.), Levang (Patrice.)," Nouvelles politiques foncières, nouveaux acteurs : des rapports fonciers sous tensions ", Territoires d'Afrique n°5. Nouvelles politiques foncières, nouveaux acteurs : des rapports fonciers sous tensions", *Territoires d'Afrique n°5.*

Nforgang(Charles.), *"Chinois au Cameroun : une incompréhension foncière",* Yaoundé, Syfa, January 2010.

Nguiffo (Samuel.),&Schwartz (Brendan),*<<The thirteenth labor of Heracles? A study of the SGSOC land concession in southwest Cameroon",*2012.

Tagne (Jean-Bruno.), *"Enquête sur la riziculture chinoise à Nanga-Eboko"*, August 2010.

Books, articles and reports on land legislation in Cameroon

Binet (Jacques.), *Droit foncier coutumier au Cameroun,* Paris, 1951, 27p.

Nyama *(Jean-Marie.),Régime foncier et domanialité publique au Cameroun,* UCAC University Press, 2012,392p.

Liz Alden (Wily)," *Whose land is it? The status of customary land ownership in Cameroon*", Ed.fenton, February 2011

Ministère des domaines, du cadastre et des affaires foncières, *Régime foncier et domanial au Cameroun : Lois et Ordonnances, Décrets et Arrêtés, Circulaires et Instructions,* January 2008.

Nguiffo (Samuel.), Mballa (Nadine.), "Les dispositions constitutionnelles, législatives et administratives relatives aux populations autochtones au Cameroun". *Les dispositions constitutionnelles, législatives et administratives relatives aux populations autochtones au Cameroun"*, 2009.

Nguiffo (Samuel.), Kenfack (Pierre-Etienne.), Mballa (Nadine.), "*L'incidence des lois foncières historiques et modernes sur les droits fonciers des communautés locales et autochtones du Cameroun*", in *les droits fonciers et les peuples des forêts d'Afrique: perspectives historiques, juridiques et anthropologiques,* January 2009.

Nguema Ondo Obiang (Samuel.), Puepi (Bernard.), "*La problématique foncière dans les pays d'Afrique centrale : Cas du Cameroun et du Gabon*", 2011.

Schwartz (Brendan.), Hoyle (David.), Nguiffo (Samuel.), "*Emerging trends in land-use conflicts in Cameroon: overlapping natural resource licenses and threats to protected areas and foreign direct investment*", June 2012.

Tchapmegni (Robinson.), "*La problématique de la propriété foncière au Cameroun*", Mbalmayo, 2005, 142.p

Books, articles and reports on China-Africa cooperation

Works

Adama (Gaye.), *Chine-Afrique : le dragon et l'autruche,* Paris, L'Harmattan, 2006,298p.

Alternative sud, *La Chine en Afrique menace ou opportunité pour le développement,* Paris, Syllepse, 2011, 184p.

Bangui (Thierry.),*Les enjeux de la coopération Chine-Afrique*, Paris L'Harmattan, 2009.

Braud (Pierre-Antoine.),*La Chine en Afrique : Anatomie d'une nouvelle stratégie chinoise*, Paris, Analysis, October 2005.

Banyonguen (Serges.),.*Rôle et responsabilité des acteurs africains dans les relations sino-africaines : Ethnographie et sociogenèse des stratégies de réceptivité,* Paris, L'Harmattan, 2013, 402p.

Boillot(Jean-Joseph.), & Dembiski *(StanislasfCHIANDIAFRIQUE: la Chine, l'Inde et l'Afrique feront le monde demain,* Odile Jacob, 2013, 370p. Chaponnière (Jean-Raphael.),etGabas (Jean-Jacques.),*Le temps de la Chine en Afrique. Enjeux et réalités au sud du Sahara,* Paris, Karthala, Collection Gemdev, 2012,216p.

Jolly (Jean.), *Les chinois à la conquête de l'Afrique,* Paris, Pygmalion, 2001, 336p.

Michel (Serge.),& Beuret (Michel.), *La Chinafrique. Pékin à la conquête du continent noir,* new expanded edition, Paris, Fayard/Pluriel, 2010, 410p.

Mbabia (Olivier.), *La Chine en Afrique : histoire géopolitique, géoéconomique,* Paris, Ellipse 2012, 157p.

Nguyen (Eric.),*China-Africa relations. L'empire du milieu à la conquête du continent noir*, Lavallois-Perret, Studyrama Perspectives 2009.

Richer *(Philippe.),L'offensive chinoise en Afrique,* Paris, Karthala, 2008, 168p.

Richer (Philippe.),*L'Afrique des chinois,* Paris, Karthala, 2012, 192p

Swieland (Tanguy Struye de.)*, La Chine et les grandes puissances en Afrique : Une approche géostratégique et géoéconomique* , Louvain, Presses Universitaires, 2010, 194p.

Articles and reports

Adama (Gueye.), "*La nouvelle donne chinoise en Afrique*", Gabriel Péri Foundation, January 2008.

Agence Française pour le Développement, ID4D conference-"*China-Africa relations: impacts for the African continent and prospects*", February 2013

Barrat (Jacques.), "*CHINAFRIQUE: Un tigre de paper",Géostratégique* n°25, October 2009

Boulan, "*Chine-Afrique : la grande désinformation*",2014.

Chaponnière (Jean-Raphael.), *"L'aide chinoise à l'Afrique : origines et enjeux*", l'Economie politique, Quarterly April 2008.

Chaponnière (Jean-Raphael.), "*Un demi siècle de relations China-Afrique*" Evolution des analyses, *Afrique contemporaine*, 2008/4 n°228

Delcourt (Laurent.), "*La Chine en Afrique : avantages ou inconvénients pour le développement?*", Alternative Sud, April 2008.

Gabas (Jean-Jacques.), "*Les investissements chinois en Afrique : à quel prix?*", in *Alternatives Internationales,* n°005, November 2007

Gwet (Guy.), "*China's power strategy as seen from Cameroon*", www.diplostratégies.blogpost.com.

Ka Mana, *"Chine-Afrique : les enjeux d'une coopération"*, in Congo-Afrique, n°425, 2008.

Kernen (Antoine.), *"Les stratégies chinoises en Afrique : du pétrole aux bassines en plastique"*, *Politique africaine,* n°105, March 2007.

Kersting (philippe.),*" Is China a major player in land grabbing in Africa?",* 2014.

Lafargue (François.),*" China and Africa La Chine et l'Afrique, perspectives chinoises"*, July-August 2005

Marchal (Roland.), *"China-Africa"*, *africulture,* n°66, March 2006

Niquet (Valérie.), "China's African strategy La stratégie africaine de la Chine", *Politique étrangère,* n°2 2006

Pairault (Thierry.), "*Les Chiffres de l'investissement direct chinois en Afrique*", *Dounia* n°3 September 2010.

Pougala (Jean-Paul.), "*The biggest lies about cooperation between China and Africa*", 2013.

Wang (Jang-Ye.), & Bio-Tchané (Abdoulaye.), "*Afrique-Chine : des liens. How to make the most of China's growing economic involvement in Africa*", *Finance &développement,* March 2008

Books, articles and reports on China

Works

Aglietta (Yves.)*, La Chine vers la superpuissance,* Paris, Economica, 2007.

Cabestan (J.-P.),*La politique internationale de la Chine,* Paris, Presses de Sciences po, 2010, 460p

Chauprade (Aymeric.), *Défis chinois: introduction à une géopolitique de la Chine,* Paris, Ellipses, 2006, 287p.

Chen (Yan.),*L'éveil de la Chine,* La Tour d'Aigues, Paris, L'Aube, 2002.

Courmont (Barthelémy.),*Chine, la grande séduction.* Essai sur le soft power chinois, Paris, Choiseul, 1990.

Cohen (philippe.),etRichard (Luc.), Le vampire du milieu : comment la Chine nous dicte sa loi, Paris, Fayard/Mille et une nuits, 2010, 336p.

Fur (Alexandre.), Gentelle (Pierre.), Pairault Chierry.),*Economie et régions de la Chine*, Paris, Armand Colin, 1999, 176.

Haber (Daniel.), Mandelbaum (Jean.),*La revanche du monde chinois?* 2^{ieme} édition, Paris, Economica, 1999, 198p.

Izraelewicz *(Eè),L'arrogance chinoise* ,Paris, Grasset, 2011, 256p

Lorot (Pascal.),*Le siècle de la Chine: Essai sur la nouvelle puissancechinoise*, Paris, Choiseul, 2007, 263p.

Ma Mung (Emmanuel.), La diaspora chinoise géographie d'une immigration, Paris, Ophrys, 2000.

Mangin (Marc.), Chine l'empire pollueur, Paris, Arthaud, 2008, 191p.

Mandelbaum (Jean.), et Haber (Daniel.),*La victoire de la Chine : L'occidentpiégépar la mondialisation,* Paris, Descartes& Cie, 2001, 136p

Articles and reports

Albertini(Dominique.), "Les terres rares, un élément majeur delapuissancechinoise"
,http://www.liberation.fr/economie/2012/03/14/les -terres-rares-un
Nhu-Nguyen Ngo, "Chine: bilan social contrasté d'un formidable essor", July 2006.

Ruoen (Ren.),*Zes performances économiques de la Chine dans le contexte international,* O.C.D.E, 1997, 180p.

Books on Africa

Bayart (Jean-François.), *L'Etat en Afrique,* Paris, Fayard, 1989, 444p.

Dambisa (Moyo.), *L'Aide fatale : Les ravages d'une aide inutile et de nouvelles solutions pour l'Afrique,* Paris, Jean-Claude Lattès, 2009, 250p Doumbia (S.),*Le manifeste pour l'Afrique : pourquoi le continent noir souffre-t-il?,* Paris, L'Harmattan, 2009.

Dussey *(Robert.),L'Afrique malade de ses hommes politiques : inconscience, irresponsabilité, ignorance ou innocence* , Paris, Jean-Picollec, 2008,252p

Esse *(Amouzou.),Pourquoi la pauvreté s'aggrave-t-elle en Afrique noire?* Paris, L'Harmattan, 2009, 220p

Courade *(Georges.),L'Afrique des idées reçues,* Paris,Belin, 2006, 400p Kabou (Axelle.),*Et si l'Afrique refusait le développement?,* Paris, L'Harmattan, 2000,207p

Mouandjo (Pierre.), Lewis (B.), *Facteurs de développement en Afrique : L'économie politique de l'Afrique au XXIème siècle, Tome II,* Paris, L'Harmattan, 2002, 358p

Osenga Babibake(Thérèse.),*Pouvoir des Organisations internationales et souveraineté des Etats ; le cas de l'union africaine,* Paris, L'Harmattan, 2010.

Pandjo Boumba (Luc.),*fa violence du développement : pouvoir, politique et rationalité économique des élites africaines,* Paris, L'Harmattan, 2002, 204p

Ray (Olivier.), and Severino (Jean-Michel.),*Le temps de l'Afrique,* Paris, Odile Jacob, 2010, 345p

Traoré (Aminata.),*L'Afrique humiliée,* Paris, Fayard, 2008, 294p

Schuerkens (Ulrike.), *Du Togo Allemand aux Togo et Ghana indépendants : changement social sous régime colonial,* Paris, Collection Etudes Africaines, L'harmattan, 2001, 624 p.

Schuerkens (Ulrike.),*Le développement social en Afrique contemporaine : une perspective de recherche inter - et intra sociétale,* Paris, L'harmattan, 1995, 174p.

Methodological books and articles

Works

Becker (Howard.),*Les ficelles du métier : comment conduire sa recherche en sciences sociales*, Paris, La découverte, Guides Repères, 2003, 352p

Beaud (Stéphane.), and Weber (Florence.), *Guide de l'enquête de terrain,* fourth edition, Paris, La Découverte,2010, 335p.

Blanchet (Alain.),& Gotman (Anne.), L'enquête est ses méthodes : L'entretien 2ième edition, Armand Colin, 2010, 128p

Boltanski (Luc.), *Zes cadres : La formation d'un groupe social,* Minuit, Paris, 1982, 523p

Brechon (Pierre.), *Enquête qualitative, enquête quantitative, Grenoble, Presses universitaires,* 2011,232p

Cefaï (Daniel.),*L'enquête de terrain,* Paris, La Découverte,

2003,624p Cohen (Samy.),*L'art d'interviewer les dirigeants,* P.U.F, Paris, 1999,277p Coenen-Huthier (Jacques.), *Observation participante et théorie sociologique,* Paris, L'Hramattan, 1995, 192p Copans *(Jean.),L'enquête et ses méthodes : L'enquête ethnologique de terrain,* Paris, Armand Colin, 2005, 126 p.

Denzin (Norman.)& Lincoln (Yvonna.), *Handbook of qualitative research.* Sage Publications, Thousand Oaks, London and New Delhi.

Durkheim (Emile.), *The Rules of Sociological Method,* Paris, P.U.F, 2007.

Grawitz (Madeleine.), *Méthodes des sciences sociales*, Paris, Dalloz, 2001,1019p

Kaufmann (Jean-Claude.),*L'enquête et ses méthodes : l'enquête compréhensif,* third edition, Paris, Armand Colin, 2011.

Loubet del Bayle (Jean-Louis.),*Ininiation aux méthodes des sciences sociales,* Paris, L'Harmattan, 2000, 272p

Mandras (Henri.), and Oberti (Marco.), *Le sociologue et son terrain : trente recherches exemplaires*, Paris, Armand Colin, 2000, 294p.

Ogien (Albert.), *Les règles de la pratique sociologique,* France, Presses universitaires, 2007, 291p

Papinot (Christian.), *Larelation d'enquête comme relation sociale. Epistémologie de la démarche de recherche ethnographique, Paris, Hermann, Collections méthodes de recherche en sciences*, 2014, 266p

Paugam (Serge.), *L'enquête sociologique,* France, Presses Universitaires, 2010,429p

Articles

Cosmovici (Ion.), "The psychology of interviewing elites", 2012, www.

ionco smovici .fr

Dayer (Caroline.), & Chamillot (Maryvonne.), "La démarche compréhensive comme moyen de construire une identité de la recherche dans les institutions de formation", Suisse, Université de Genène, n° 14, 2012.

Dantier (Bernard.), "La chose sociologique et sa représentation : Introduction aux règles de la méthode sociologique d'Emile Durkheim (1895)", Classique de sciences sociales. Chicoutimi, January 1st2003,53pwww.classique.uqac.ca/contemporain/.../Dantier intr règ les_methode.doc

Lefebvre (Nicolas.), "L'entretien comme méthode de recherche",www.staps.univ_lille2.fr/fileadmin/user_upload/.../entre_meth_recherche.pdf.

Olivier de Sardan (Jean-Pierre.), "L'enquête socio-anthropologique de terrain : synthèse méthodologique à usage des étudiants", études et travaux, no. 13, October 2003.

Projet Base " Apprentissage des notions de base en sciences économiques : Max Weber et la sociologie compréhensive ", Lausane, University, www.idéal-type,sociologie moderne

Yon (Guillaume.), "Introduction à la socio-histoire de Gérard Noiriel", www2 .cndp.fr/revuesdeess/notelecture/2006.9-14.htm

Theses and dissertations

Theses

Belomo Essono (Pélagie Chantal.), L'ordre et la sécurité publics dans la construction de l'Etat au Cameroun, Université Montesquieu-Bordeaux IV, Thèse en science politique, 2007.

Hatcheu Tchawe (Emile.), L'approvisionnement et la distribution alimentaire à Douala (Cameroun) : logiques sociales et pratiques spatiales des acteurs, Université Paris 1, Thèse en Géographie, 2003.

Issoufou (Oumarou.), "Femmes et développement local: Analyse socio-anthropologique de l'organisation foncière au Niger. Le cas de la région de Tillabery, Université de Rennes 2- Haute Bretagne, Thèse de sociologie, 2008.

Tchapmegni (Robinson.), Le contentieux de la propriété foncière au Cameroun, Université de Nantes, Thèse de droit privé, 2008.

Memories

Bessala Mviani (Alvine.), La percée des investissements privés chinois au Cameroun de 1972 à 2004, Master's thesis, University of Yaoundé 1, 2004

Pial Pial (Jean-Claude.), Le pluralisme juridique dans la gestion foncière dans l'arrondissement de Doume (EST-Cameroun) : Rôle et responsabilité de l'animateur, Mémoire de Conseiller Principal de Jeunesse et d'animation, 2012.

Tchambia Paho (Landry Fulbert.), Contribution à la mise en place d'une démarche qualité à la société camerounaise de palmeraies Socapalm, DESS thesis in quality control and management, ENSAI, Ngaoundéré, 2005.

I want morebooks!

Buy your books fast and straightforward online - at one of world's fastest growing online book stores! Environmentally sound due to Print-on-Demand technologies.

Buy your books online at
www.morebooks.shop

Kaufen Sie Ihre Bücher schnell und unkompliziert online – auf einer der am schnellsten wachsenden Buchhandelsplattformen weltweit! Dank Print-On-Demand umwelt- und ressourcenschonend produziert.

Bücher schneller online kaufen
www.morebooks.shop

Printed by Books on Demand GmbH, Norderstedt / Germany